KB174523

PILOT FLIGHT LOGBOOK

Logbook Number _____

Period _____ ~ _____

PILOT'S NAME _____

PERMANENT MAILING ADDRESS _____

CHANGE OF ADDRESS _____

E-MAIL / MOBILE _____

RECORD OF CERTIFICATES and RATINGS

CERTIFICATES			RATINGS		
TYPE	NUMBER	DATE OF ISSUE	CATEGORY, CLASS, OR TYPE	DATE OF ISSUE	TOTAL TIME

AIRCRAFT RECORD BY MODEL																FLIGHT PROFICIENCY				
MODEL _____ YEAR _____					MODEL _____ YEAR _____					MODEL _____ YEAR _____						RECURRENT CHECK	PIC PROFICIENCY CHECK	LIP RECURRENT TRAINING	CFI RENEWAL	
	PIC	SIC	IP			PIC	SIC	IP			PIC	SIC	IP							

DATE (MON/DAY)	AIRCRAFT TYPE	AIRCRAFT IDENT	ROUTE OF FLIGHT		FLIGHT NUMBER	TAKE OFF & LANDING			CONDITION OF FLIGHT				FLIG SIMUL.
			FROM	TO		DAY	NIGHT	AUTO LAND	NIGHT	ACTUAL INSTRUMENT	APP		
											R/W NO.	TYPE	
						/	/						
						/	/						
						/	/						
						/	/						
						/	/						
						/	/						
						/	/						
						/	/						
						/	/						
						/	/						
						/	/						
THIS RECORD IS CERTIFIED TRUE AND CORRECT PILOT'S SIGNATURE _____			PAGE TOTAL			/	/						
			PREVIOUS TOTAL			/	/						
			NEW TOTAL			/	/						

TYPE OF PILOTING TIME						REMARKS AND ENDORSEMENTS
PIC	SIC	AS FLIGHT INSTRUCTOR	DUAL RECEIVED			

DATE (MON/DAY)	AIRCRAFT TYPE	AIRCRAFT IDENT	ROUTE OF FLIGHT		FLIGHT NUMBER	TAKE OFF & LANDING			CONDITION OF FLIGHT				FLIG SIMUL
			FROM	TO		DAY	NIGHT	AUTO LAND	NIGHT	ACTUAL INSTRUMENT	APP		
											R/W NO.	TYPE	
						/	/						
						/	/						
						/	/						
						/	/						
						/	/						
						/	/						
						/	/						
						/	/						
						/	/						
						/	/						
						/	/						
THIS RECORD IS CERTIFIED TRUE AND CORRECT PILOT'S SIGNATURE _____			PAGE TOTAL			/	/						
			PREVIOUS TOTAL			/	/						
			NEW TOTAL			/	/						

TYPE OF PILOTING TIME						REMARKS AND ENDORSEMENTS
PIC	SIC	AS FLIGHT INSTRUCTOR	DUAL RECEIVED			

DATE (MON/DAY)	AIRCRAFT TYPE	AIRCRAFT IDENT	ROUTE OF FLIGHT		FLIGHT NUMBER	TAKE OFF & LANDING			CONDITION OF FLIGHT				FLIG SIMUL
			FROM	TO		DAY	NIGHT	AUTO LAND	NIGHT	ACTUAL INSTRUMENT	APP		
											R/W NO.	TYPE	
						/	/						
						/	/						
						/	/						
						/	/						
						/	/						
						/	/						
						/	/						
						/	/						
						/	/						
						/	/						
						/	/						
THIS RECORD IS CERTIFIED TRUE AND CORRECT PILOT'S SIGNATURE _____			PAGE TOTAL			/	/						
			PREVIOUS TOTAL			/	/						
			NEW TOTAL			/	/						

		TYPE OF PILOTING TIME				REMARKS AND ENDORSEMENTS
PIC	SIC	AS FLIGHT INSTRUCTOR	DUAL RECEIVED			

DATE (MON/DAY)	AIRCRAFT TYPE	AIRCRAFT IDENT	ROUTE OF FLIGHT		FLIGHT NUMBER	TAKE OFF & LANDING			CONDITION OF FLIGHT				FLIG SIMULA
			FROM	TO		DAY	NIGHT	AUTO LAND	NIGHT	ACTUAL INSTRUMENT	R/W NO.	TYPE	
						/	/						
						/	/						
						/	/						
						/	/						
						/	/						
						/	/						
						/	/						
						/	/						
						/	/						
						/	/						
						/	/						
THIS RECORD IS CERTIFIED TRUE AND CORRECT PILOT'S SIGNATURE _____			PAGE TOTAL			/	/						
			PREVIOUS TOTAL			/	/						
			NEW TOTAL			/	/						

TYPE OF PILOTING TIME						REMARKS AND ENDORSEMENTS
PIC	SIC	AS FLIGHT INSTRUCTOR	DUAL RECEIVED			

DATE (MON/DAY)	AIRCRAFT TYPE	AIRCRAFT IDENT	ROUTE OF FLIGHT		FLIGHT NUMBER	TAKE OFF & LANDING			CONDITION OF FLIGHT				FLIGHT SIMULA
			FROM	TO		DAY	NIGHT	AUTO LAND	NIGHT	ACTUAL INSTRUMENT	APP		
											R/W NO.	TYPE	
						/	/						
						/	/						
						/	/						
						/	/						
						/	/						
						/	/						
						/	/						
						/	/						
						/	/						
						/	/						
						/	/						
THIS RECORD IS CERTIFIED TRUE AND CORRECT PILOT'S SIGNATURE _____			PAGE TOTAL			/	/						
			PREVIOUS TOTAL			/	/						
			NEW TOTAL			/	/						

TYPE OF PILOTING TIME						REMARKS AND ENDORSEMENTS
PIC	SIC	AS FLIGHT INSTRUCTOR	DUAL RECEIVED			

DATE (MON/DAY)	AIRCRAFT TYPE	AIRCRAFT IDENT	ROUTE OF FLIGHT		FLIGHT NUMBER	TAKE OFF & LANDING			CONDITION OF FLIGHT				FLIG SIMUL
			FROM	TO		DAY	NIGHT	AUTO LAND	NIGHT	ACTUAL INSTRUMENT	APP		
											R/W NO.	TYPE	
						/	/						
						/	/						
						/	/						
						/	/						
						/	/						
						/	/						
						/	/						
						/	/						
						/	/						
						/	/						
						/	/						

THIS RECORD IS CERTIFIED TRUE AND CORRECT PILOT'S SIGNATURE _____			PAGE TOTAL		/	/					
			PREVIOUS TOTAL		/	/					
			NEW TOTAL		/	/					

TYPE OF PILOTING TIME						REMARKS AND ENDORSEMENTS
PIC	SIC	AS FLIGHT INSTRUCTOR	DUAL RECEIVED			

DATE (MON/DAY)	AIRCRAFT TYPE	AIRCRAFT IDENT	ROUTE OF FLIGHT		FLIGHT NUMBER	TAKE OFF & LANDING			CONDITION OF FLIGHT				FLIG SIMUL
			FROM	TO		DAY	NIGHT	AUTO LAND	NIGHT	ACTUAL INSTRUMENT	APP		
											R/W NO.	TYPE	
						/	/						
						/	/						
						/	/						
						/	/						
						/	/						
						/	/						
						/	/						
						/	/						
						/	/						
						/	/						
						/	/						
THIS RECORD IS CERTIFIED TRUE AND CORRECT PILOT'S SIGNATURE _____			PAGE TOTAL			/	/						
			PREVIOUS TOTAL			/	/						
			NEW TOTAL			/	/						

TYPE OF PILOTING TIME						REMARKS AND ENDORSEMENTS
PIC	SIC	AS FLIGHT INSTRUCTOR	DUAL RECEIVED			

DATE (MON/DAY)	AIRCRAFT TYPE	AIRCRAFT IDENT	ROUTE OF FLIGHT		FLIGHT NUMBER	TAKE OFF & LANDING			CONDITION OF FLIGHT				FLIG SIMUL/
			FROM	TO		DAY	NIGHT	AUTO LAND	NIGHT	ACTUAL INSTRUMENT	APP		
											R/W NO.	TYPE	
						/	/						
						/	/						
						/	/						
						/	/						
						/	/						
						/	/						
						/	/						
						/	/						
						/	/						
						/	/						
						/	/						
THIS RECORD IS CERTIFIED TRUE AND CORRECT PILOT'S SIGNATURE _____			PAGE TOTAL			/	/						
			PREVIOUS TOTAL			/	/						
			NEW TOTAL			/	/						

TYPE OF PILOTING TIME						REMARKS AND ENDORSEMENTS
PIC	SIC	AS FLIGHT INSTRUCTOR	DUAL RECEIVED			

DATE (MON/DAY)	AIRCRAFT TYPE	AIRCRAFT IDENT	ROUTE OF FLIGHT		FLIGHT NUMBER	TAKE OFF & LANDING			CONDITION OF FLIGHT				FLIG SIMUL/
			FROM	TO		DAY	NIGHT	AUTO LAND	NIGHT	ACTUAL INSTRUMENT	APP		
											R/W NO.	TYPE	
						/	/						
						/	/						
						/	/						
						/	/						
						/	/						
						/	/						
						/	/						
						/	/						
						/	/						
						/	/						
						/	/						
THIS RECORD IS CERTIFIED TRUE AND CORRECT PILOT'S SIGNATURE _____			PAGE TOTAL			/	/						
			PREVIOUS TOTAL			/	/						
			NEW TOTAL			/	/						

		TYPE OF PILOTING TIME				REMARKS AND ENDORSEMENTS
PIC	SIC	AS FLIGHT INSTRUCTOR	DUAL RECEIVED			

20 _____

DATE (MON/DAY)	AIRCRAFT TYPE	AIRCRAFT IDENT	ROUTE OF FLIGHT		FLIGHT NUMBER	TAKE OFF & LANDING			CONDITION OF FLIGHT				FLIG SIMUL
			FROM	TO		DAY	NIGHT	AUTO LAND	NIGHT	ACTUAL INSTRUMENT	APP		
											R/W NO.	TYPE	
						/	/						
						/	/						
						/	/						
						/	/						
						/	/						
						/	/						
						/	/						
						/	/						
						/	/						
						/	/						
						/	/						

THIS RECORD IS CERTIFIED TRUE AND CORRECT PILOT'S SIGNATURE _____			PAGE TOTAL	/	/						
			PREVIOUS TOTAL	/	/						
			NEW TOTAL	/	/						

		TYPE OF PILOTING TIME				REMARKS AND ENDORSEMENTS
PIC	SIC	AS FLIGHT INSTRUCTOR	DUAL RECEIVED			

| DATE (MON/DAY) | AIRCRAFT TYPE | AIRCRAFT IDENT | ROUTE OF FLIGHT | | FLIGHT NUMBER | TAKE OFF & LANDING | | | CONDITION OF FLIGHT | | | | FLIC SIMUL |
| | | | FROM | TO | | DAY | NIGHT | AUTO LAND | NIGHT | ACTUAL INSTRUMENT | APP | | |
											R/W NO.	TYPE	
						/	/						
						/	/						
						/	/						
						/	/						
						/	/						
						/	/						
						/	/						
						/	/						
						/	/						
						/	/						
						/	/						

THIS RECORD IS CERTIFIED TRUE AND CORRECT PILOT'S SIGNATURE _____

PAGE TOTAL	/	/
PREVIOUS TOTAL	/	/
NEW TOTAL	/	/

TYPE OF PILOTING TIME						REMARKS AND ENDORSEMENTS
PIC	SIC	AS FLIGHT INSTRUCTOR	DUAL RECEIVED			

DATE (MON/DAY)	AIRCRAFT TYPE	AIRCRAFT IDENT	ROUTE OF FLIGHT		FLIGHT NUMBER	TAKE OFF & LANDING			CONDITION OF FLIGHT				FLIG SIMUL
			FROM	TO		DAY	NIGHT	AUTO LAND	NIGHT	ACTUAL INSTRUMENT	APP		
											R/W NO.	TYPE	
						/	/						
						/	/						
						/	/						
						/	/						
						/	/						
						/	/						
						/	/						
						/	/						
						/	/						
						/	/						
						/	/						

THIS RECORD IS CERTIFIED TRUE AND CORRECT PILOT'S SIGNATURE _____	PAGE TOTAL	/	/					
	PREVIOUS TOTAL	/	/					
	NEW TOTAL	/	/					

TYPE OF PILOTING TIME						REMARKS AND ENDORSEMENTS
PIC	SIC	AS FLIGHT INSTRUCTOR	DUAL RECEIVED			

DATE (MON/DAY)	AIRCRAFT TYPE	AIRCRAFT IDENT	ROUTE OF FLIGHT		FLIGHT NUMBER	TAKE OFF & LANDING			CONDITION OF FLIGHT				FLIG SIMUL.
			FROM	TO		DAY	NIGHT	AUTO LAND	NIGHT	ACTUAL INSTRUMENT	APP		
											R/W NO.	TYPE	
						/	/						
						/	/						
						/	/						
						/	/						
						/	/						
						/	/						
						/	/						
						/	/						
						/	/						
						/	/						
						/	/						
THIS RECORD IS CERTIFIED TRUE AND CORRECT PILOT'S SIGNATURE _____			PAGE TOTAL			/	/						
			PREVIOUS TOTAL			/	/						
			NEW TOTAL			/	/						

TYPE OF PILOTING TIME						REMARKS AND ENDORSEMENTS
PIC	SIC	AS FLIGHT INSTRUCTOR	DUAL RECEIVED			

DATE (MON/DAY)	AIRCRAFT TYPE	AIRCRAFT IDENT	ROUTE OF FLIGHT		FLIGHT NUMBER	TAKE OFF & LANDING			CONDITION OF FLIGHT				FLIG SIMULA
			FROM	TO		DAY	NIGHT	AUTO LAND	NIGHT	ACTUAL INSTRUMENT	APP		
											R/W NO.	TYPE	
						/	/						
						/	/						
						/	/						
						/	/						
						/	/						
						/	/						
						/	/						
						/	/						
						/	/						
						/	/						
						/	/						
THIS RECORD IS CERTIFIED TRUE AND CORRECT PILOT'S SIGNATURE _____			PAGE TOTAL			/	/						
			PREVIOUS TOTAL			/	/						
			NEW TOTAL			/	/						

TYPE OF PILOTING TIME						REMARKS AND ENDORSEMENTS
PIC	SIC	AS FLIGHT INSTRUCTOR	DUAL RECEIVED			

DATE (MON/DAY)	AIRCRAFT TYPE	AIRCRAFT IDENT	ROUTE OF FLIGHT		FLIGHT NUMBER	TAKE OFF & LANDING			CONDITION OF FLIGHT				FLIG SIMUL
			FROM	TO		DAY	NIGHT	AUTO LAND	NIGHT	ACTUAL INSTRUMENT	APP		
											R/W NO.	TYPE	
						/	/						
						/	/						
						/	/						
						/	/						
						/	/						
						/	/						
						/	/						
						/	/						
						/	/						
						/	/						
						/	/						
THIS RECORD IS CERTIFIED TRUE AND CORRECT PILOT'S SIGNATURE _____			PAGE TOTAL			/	/						
			PREVIOUS TOTAL			/	/						
			NEW TOTAL			/	/						

TYPE OF PILOTING TIME						REMARKS AND ENDORSEMENTS
PIC	SIC	AS FLIGHT INSTRUCTOR	DUAL RECEIVED			

DATE (MON/DAY)	AIRCRAFT TYPE	AIRCRAFT IDENT	ROUTE OF FLIGHT		FLIGHT NUMBER	TAKE OFF & LANDING			CONDITION OF FLIGHT				FLIG SIMUL
			FROM	TO		DAY	NIGHT	AUTO LAND	NIGHT	ACTUAL INSTRUMENT	R/W NO.	TYPE	
						/	/						
						/	/						
						/	/						
						/	/						
						/	/						
						/	/						
						/	/						
						/	/						
						/	/						
						/	/						
						/	/						
THIS RECORD IS CERTIFIED TRUE AND CORRECT PILOT'S SIGNATURE _____			PAGE TOTAL			/	/						
			PREVIOUS TOTAL			/	/						
			NEW TOTAL			/	/						

TYPE OF PILOTING TIME						REMARKS AND ENDORSEMENTS
PIC	SIC	AS FLIGHT INSTRUCTOR	DUAL RECEIVED			

20 _____

DATE (MON/DAY)	AIRCRAFT TYPE	AIRCRAFT IDENT	ROUTE OF FLIGHT		FLIGHT NUMBER	TAKE OFF & LANDING			CONDITION OF FLIGHT				FLIG SIMUL
			FROM	TO		DAY	NIGHT	AUTO LAND	NIGHT	ACTUAL INSTRUMENT	APP R/W NO.	TYPE	
						/	/						
						/	/						
						/	/						
						/	/						
						/	/						
						/	/						
						/	/						
						/	/						
						/	/						
						/	/						
						/	/						
THIS RECORD IS CERTIFIED TRUE AND CORRECT PILOT'S SIGNATURE _____			PAGE TOTAL			/	/						
			PREVIOUS TOTAL			/	/						
			NEW TOTAL			/	/						

TYPE OF PILOTING TIME						REMARKS AND ENDORSEMENTS
PIC	SIC	AS FLIGHT INSTRUCTOR	DUAL RECEIVED			

DATE (MON/DAY)	AIRCRAFT TYPE	AIRCRAFT IDENT	ROUTE OF FLIGHT		FLIGHT NUMBER	TAKE OFF & LANDING			CONDITION OF FLIGHT				FLIGHT SIMULA
			FROM	TO		DAY	NIGHT	AUTO LAND	NIGHT	ACTUAL INSTRUMENT	APP		
											R/W NO.	TYPE	
						/	/						
						/	/						
						/	/						
						/	/						
						/	/						
						/	/						
						/	/						
						/	/						
						/	/						
						/	/						
						/	/						
THIS RECORD IS CERTIFIED TRUE AND CORRECT PILOT'S SIGNATURE _____			PAGE TOTAL			/	/						
			PREVIOUS TOTAL			/	/						
			NEW TOTAL			/	/						

TYPE OF PILOTING TIME						REMARKS AND ENDORSEMENTS
PIC	SIC	AS FLIGHT INSTRUCTOR	DUAL RECEIVED			

DATE (MON/DAY)	AIRCRAFT TYPE	AIRCRAFT IDENT	ROUTE OF FLIGHT		FLIGHT NUMBER	TAKE OFF & LANDING			CONDITION OF FLIGHT				FLIG SIMUL
			FROM	TO		DAY	NIGHT	AUTO LAND	NIGHT	ACTUAL INSTRUMENT	APP		
											R/W NO.	TYPE	
						/	/						
						/	/						
						/	/						
						/	/						
						/	/						
						/	/						
						/	/						
						/	/						
						/	/						
						/	/						
						/	/						
THIS RECORD IS CERTIFIED TRUE AND CORRECT PILOT'S SIGNATURE _____			PAGE TOTAL			/	/						
			PREVIOUS TOTAL			/	/						
			NEW TOTAL			/	/						

TYPE OF PILOTING TIME						REMARKS AND ENDORSEMENTS
PIC	SIC	AS FLIGHT INSTRUCTOR	DUAL RECEIVED			

DATE (MON/DAY)	AIRCRAFT TYPE	AIRCRAFT IDENT	ROUTE OF FLIGHT		FLIGHT NUMBER	TAKE OFF & LANDING			CONDITION OF FLIGHT				FLIG SIMUL
			FROM	TO		DAY	NIGHT	AUTO LAND	NIGHT	ACTUAL INSTRUMENT	APP		
											R/W NO.	TYPE	
						/	/						
						/	/						
						/	/						
						/	/						
						/	/						
						/	/						
						/	/						
						/	/						
						/	/						
						/	/						
						/	/						

THIS RECORD IS CERTIFIED TRUE AND CORRECT PILOT'S SIGNATURE _____	PAGE TOTAL	/	/					
	PREVIOUS TOTAL	/	/					
	NEW TOTAL	/	/					

TYPE OF PILOTING TIME						REMARKS AND ENDORSEMENTS
PIC	SIC	AS FLIGHT INSTRUCTOR	DUAL RECEIVED			

DATE (MON/DAY)	AIRCRAFT TYPE	AIRCRAFT IDENT	ROUTE OF FLIGHT		FLIGHT NUMBER	TAKE OFF & LANDING			CONDITION OF FLIGHT				FLIG SIMULA
			FROM	TO		DAY	NIGHT	AUTO LAND	NIGHT	ACTUAL INSTRUMENT	APP		
											R/W NO.	TYPE	
						/	/						
						/	/						
						/	/						
						/	/						
						/	/						
						/	/						
						/	/						
						/	/						
						/	/						
						/	/						
						/	/						
THIS RECORD IS CERTIFIED TRUE AND CORRECT PILOT'S SIGNATURE _____			PAGE TOTAL			/	/						
			PREVIOUS TOTAL			/	/						
			NEW TOTAL			/	/						

TYPE OF PILOTING TIME						REMARKS AND ENDORSEMENTS
PIC	SIC	AS FLIGHT INSTRUCTOR	DUAL RECEIVED			

DATE (MON/DAY)	AIRCRAFT TYPE	AIRCRAFT IDENT	ROUTE OF FLIGHT		FLIGHT NUMBER	TAKE OFF & LANDING			CONDITION OF FLIGHT				FLIG SIMUL.
			FROM	TO		DAY	NIGHT	AUTO LAND	NIGHT	ACTUAL INSTRUMENT	APP		
											R/W NO.	TYPE	
						/	/						
						/	/						
						/	/						
						/	/						
						/	/						
						/	/						
						/	/						
						/	/						
						/	/						
						/	/						
						/	/						
THIS RECORD IS CERTIFIED TRUE AND CORRECT PILOT'S SIGNATURE _____			PAGE TOTAL			/	/						
			PREVIOUS TOTAL			/	/						
			NEW TOTAL			/	/						

TYPE OF PILOTING TIME						REMARKS AND ENDORSEMENTS
PIC	SIC	AS FLIGHT INSTRUCTOR	DUAL RECEIVED			

20 _____

DATE (MON/DAY)	AIRCRAFT TYPE	AIRCRAFT IDENT	ROUTE OF FLIGHT		FLIGHT NUMBER	TAKE OFF & LANDING			CONDITION OF FLIGHT				FLIG SIMUL
			FROM	TO		DAY	NIGHT	AUTO LAND	NIGHT	ACTUAL INSTRUMENT	APP		
											R/W NO.	TYPE	
						/	/						
						/	/						
						/	/						
						/	/						
						/	/						
						/	/						
						/	/						
						/	/						
						/	/						
						/	/						
						/	/						

THIS RECORD IS CERTIFIED TRUE AND CORRECT PILOT'S SIGNATURE _____	PAGE TOTAL	/	/					
	PREVIOUS TOTAL	/	/					
	NEW TOTAL	/	/					

TYPE OF PILOTING TIME						REMARKS AND ENDORSEMENTS
PIC	SIC	AS FLIGHT INSTRUCTOR	DUAL RECEIVED			

| DATE (MON/DAY) | AIRCRAFT TYPE | AIRCRAFT IDENT | ROUTE OF FLIGHT | | FLIGHT NUMBER | TAKE OFF & LANDING | | | CONDITION OF FLIGHT | | | | FLIG SIMUL |
			FROM	TO		DAY	NIGHT	AUTO LAND	NIGHT	ACTUAL INSTRUMENT	R/W NO.	TYPE	
						/	/						
						/	/						
						/	/						
						/	/						
						/	/						
						/	/						
						/	/						
						/	/						
						/	/						
						/	/						
						/	/						
THIS RECORD IS CERTIFIED TRUE AND CORRECT PILOT'S SIGNATURE _____			PAGE TOTAL			/	/						
			PREVIOUS TOTAL			/	/						
			NEW TOTAL			/	/						

TYPE OF PILOTING TIME						REMARKS AND ENDORSEMENTS
PIC	SIC	AS FLIGHT INSTRUCTOR	DUAL RECEIVED			

DATE (MON/DAY)	AIRCRAFT TYPE	AIRCRAFT IDENT	ROUTE OF FLIGHT		FLIGHT NUMBER	TAKE OFF & LANDING			CONDITION OF FLIGHT				FLIGH SIMULA
			FROM	TO		DAY	NIGHT	AUTO LAND	NIGHT	ACTUAL INSTRUMENT	APP R/W NO.	TYPE	
						/	/						
						/	/						
						/	/						
						/	/						
						/	/						
						/	/						
						/	/						
						/	/						
						/	/						
						/	/						
						/	/						
THIS RECORD IS CERTIFIED TRUE AND CORRECT PILOT'S SIGNATURE _____			PAGE TOTAL			/	/						
			PREVIOUS TOTAL			/	/						
			NEW TOTAL			/	/						

PIC	SIC	TYPE OF PILOTING TIME				REMARKS AND ENDORSEMENTS
		AS FLIGHT INSTRUCTOR	DUAL RECEIVED			

DATE (MON/DAY)	AIRCRAFT TYPE	AIRCRAFT IDENT	ROUTE OF FLIGHT		FLIGHT NUMBER	TAKE OFF & LANDING			CONDITION OF FLIGHT				FLIG SIMUL
			FROM	TO		DAY	NIGHT	AUTO LAND	NIGHT	ACTUAL INSTRUMENT	APP		
											R/W NO.	TYPE	
						/	/						
						/	/						
						/	/						
						/	/						
						/	/						
						/	/						
						/	/						
						/	/						
						/	/						
						/	/						
						/	/						
THIS RECORD IS CERTIFIED TRUE AND CORRECT PILOT'S SIGNATURE _____			PAGE TOTAL			/	/						
			PREVIOUS TOTAL			/	/						
			NEW TOTAL			/	/						

		TYPE OF PILOTING TIME				REMARKS AND ENDORSEMENTS
PIC	SIC	AS FLIGHT INSTRUCTOR	DUAL RECEIVED			

DATE (MON/DAY)	AIRCRAFT TYPE	AIRCRAFT IDENT	ROUTE OF FLIGHT		FLIGHT NUMBER	TAKE OFF & LANDING			CONDITION OF FLIGHT				FLIC SIMUL
			FROM	TO		DAY	NIGHT	AUTO LAND	NIGHT	ACTUAL INSTRUMENT	APP		
											R/W NO.	TYPE	
						/	/						
						/	/						
						/	/						
						/	/						
						/	/						
						/	/						
						/	/						
						/	/						
						/	/						
						/	/						
						/	/						

THIS RECORD IS CERTIFIED TRUE AND CORRECT
PILOT'S
SIGNATURE _____

PAGE TOTAL	/	/
PREVIOUS TOTAL	/	/
NEW TOTAL	/	/

TYPE OF PILOTING TIME						REMARKS AND ENDORSEMENTS
PIC	SIC	AS FLIGHT INSTRUCTOR	DUAL RECEIVED			

DATE (MON/DAY)	AIRCRAFT TYPE	AIRCRAFT IDENT	ROUTE OF FLIGHT		FLIGHT NUMBER	TAKE OFF & LANDING			CONDITION OF FLIGHT				FLIG SIMUL
			FROM	TO		DAY	NIGHT	AUTO LAND	NIGHT	ACTUAL INSTRUMENT	APP		
											R/W NO.	TYPE	
						/	/						
						/	/						
						/	/						
						/	/						
						/	/						
						/	/						
						/	/						
						/	/						
						/	/						
						/	/						
						/	/						
THIS RECORD IS CERTIFIED TRUE AND CORRECT PILOT'S SIGNATURE _____			PAGE TOTAL			/	/						
			PREVIOUS TOTAL			/	/						
			NEW TOTAL			/	/						

TYPE OF PILOTING TIME						REMARKS AND ENDORSEMENTS
PIC	SIC	AS FLIGHT INSTRUCTOR	DUAL RECEIVED			

DATE (MON/DAY)	AIRCRAFT TYPE	AIRCRAFT IDENT	ROUTE OF FLIGHT		FLIGHT NUMBER	TAKE OFF & LANDING			CONDITION OF FLIGHT				FLIGHT SIMULA
			FROM	TO		DAY	NIGHT	AUTO LAND	NIGHT	ACTUAL INSTRUMENT	APP		
											R/W NO.	TYPE	
						/	/						
						/	/						
						/	/						
						/	/						
						/	/						
						/	/						
						/	/						
						/	/						
						/	/						
						/	/						
						/	/						
THIS RECORD IS CERTIFIED TRUE AND CORRECT PILOT'S SIGNATURE _____			PAGE TOTAL			/	/						
			PREVIOUS TOTAL			/	/						
			NEW TOTAL			/	/						

TYPE OF PILOTING TIME						REMARKS AND ENDORSEMENTS
PIC	SIC	AS FLIGHT INSTRUCTOR	DUAL RECEIVED			

DATE (MON/DAY)	AIRCRAFT TYPE	AIRCRAFT IDENT	ROUTE OF FLIGHT		FLIGHT NUMBER	TAKE OFF & LANDING			CONDITION OF FLIGHT				FLIG SIMULA
			FROM	TO		DAY	NIGHT	AUTO LAND	NIGHT	ACTUAL INSTRUMENT	APP R/W NO.	TYPE	
						/	/						
						/	/						
						/	/						
						/	/						
						/	/						
						/	/						
						/	/						
						/	/						
						/	/						
						/	/						
						/	/						

THIS RECORD IS CERTIFIED TRUE AND CORRECT PILOT'S SIGNATURE _____

PAGE TOTAL	/	/
PREVIOUS TOTAL	/	/
NEW TOTAL	/	/

TYPE OF PILOTING TIME						REMARKS AND ENDORSEMENTS
PIC	SIC	AS FLIGHT INSTRUCTOR	DUAL RECEIVED			

20 _____

| DATE (MON/DAY) | AIRCRAFT TYPE | AIRCRAFT IDENT | ROUTE OF FLIGHT | | FLIGHT NUMBER | TAKE OFF & LANDING | | | CONDITION OF FLIGHT | | | | FLIG SIMUL |
| | | | FROM | TO | | DAY | NIGHT | AUTO LAND | NIGHT | ACTUAL INSTRUMENT | APP | | |
											R/W NO.	TYPE	
						/	/						
						/	/						
						/	/						
						/	/						
						/	/						
						/	/						
						/	/						
						/	/						
						/	/						
						/	/						
						/	/						

THIS RECORD IS CERTIFIED TRUE AND CORRECT PILOT'S SIGNATURE _____

PAGE TOTAL	/	/
PREVIOUS TOTAL	/	/
NEW TOTAL	/	/

TYPE OF PILOTING TIME						REMARKS AND ENDORSEMENTS
PIC	SIC	AS FLIGHT INSTRUCTOR	DUAL RECEIVED			

DATE (MON/DAY)	AIRCRAFT TYPE	AIRCRAFT IDENT	ROUTE OF FLIGHT		FLIGHT NUMBER	TAKE OFF & LANDING			CONDITION OF FLIGHT				FLIGHT SIMULA
			FROM	TO		DAY	NIGHT	AUTO LAND	NIGHT	ACTUAL INSTRUMENT	APP		
											R/W NO.	TYPE	
						/	/						
						/	/						
						/	/						
						/	/						
						/	/						
						/	/						
						/	/						
						/	/						
						/	/						
						/	/						
						/	/						

THIS RECORD IS CERTIFIED TRUE AND CORRECT PILOT'S SIGNATURE _____	PAGE TOTAL	/	/					
	PREVIOUS TOTAL	/	/					
	NEW TOTAL	/	/					

TYPE OF PILOTING TIME						REMARKS AND ENDORSEMENTS
PIC	SIC	AS FLIGHT INSTRUCTOR	DUAL RECEIVED			

DATE (MON/DAY)	AIRCRAFT TYPE	AIRCRAFT IDENT	ROUTE OF FLIGHT		FLIGHT NUMBER	TAKE OFF & LANDING			CONDITION OF FLIGHT				FLIG SIMUL
			FROM	TO		DAY	NIGHT	AUTO LAND	NIGHT	ACTUAL INSTRUMENT	APP R/W NO.	TYPE	
						/	/						
						/	/						
						/	/						
						/	/						
						/	/						
						/	/						
						/	/						
						/	/						
						/	/						
						/	/						
						/	/						
THIS RECORD IS CERTIFIED TRUE AND CORRECT PILOT'S SIGNATURE _____			PAGE TOTAL			/	/						
			PREVIOUS TOTAL			/	/						
			NEW TOTAL			/	/						

TYPE OF PILOTING TIME						REMARKS AND ENDORSEMENTS
PIC	SIC	AS FLIGHT INSTRUCTOR	DUAL RECEIVED			

DATE (MON/DAY)	AIRCRAFT TYPE	AIRCRAFT IDENT	ROUTE OF FLIGHT		FLIGHT NUMBER	TAKE OFF & LANDING			CONDITION OF FLIGHT				FLIG SIMUL
			FROM	TO		DAY	NIGHT	AUTO LAND	NIGHT	ACTUAL INSTRUMENT	APP		
											R/W NO.	TYPE	
						/	/						
						/	/						
						/	/						
						/	/						
						/	/						
						/	/						
						/	/						
						/	/						
						/	/						
						/	/						
						/	/						
THIS RECORD IS CERTIFIED TRUE AND CORRECT PILOT'S SIGNATURE _____			PAGE TOTAL			/	/						
			PREVIOUS TOTAL			/	/						
			NEW TOTAL			/	/						

TYPE OF PILOTING TIME						REMARKS AND ENDORSEMENTS
PIC	SIC	AS FLIGHT INSTRUCTOR	DUAL RECEIVED			

DATE (MON/DAY)	AIRCRAFT TYPE	AIRCRAFT IDENT	ROUTE OF FLIGHT		FLIGHT NUMBER	TAKE OFF & LANDING			CONDITION OF FLIGHT				FLIG SIMUL
			FROM	TO		DAY	NIGHT	AUTO LAND	NIGHT	ACTUAL INSTRUMENT	APP		
											R/W NO.	TYPE	
						/	/						
						/	/						
						/	/						
						/	/						
						/	/						
						/	/						
						/	/						
						/	/						
						/	/						
						/	/						
						/	/						

THIS RECORD IS CERTIFIED TRUE AND CORRECT PILOT'S SIGNATURE _____			PAGE TOTAL			/	/						
			PREVIOUS TOTAL			/	/						
			NEW TOTAL			/	/						

TYPE OF PILOTING TIME						REMARKS AND ENDORSEMENTS
PIC	SIC	AS FLIGHT INSTRUCTOR	DUAL RECEIVED			

DATE (MON/DAY)	AIRCRAFT TYPE	AIRCRAFT IDENT	ROUTE OF FLIGHT		FLIGHT NUMBER	TAKE OFF & LANDING			CONDITION OF FLIGHT				FLIG SIMUL
			FROM	TO		DAY	NIGHT	AUTO LAND	NIGHT	ACTUAL INSTRUMENT	APP		
											R/W NO.	TYPE	
						/	/						
						/	/						
						/	/						
						/	/						
						/	/						
						/	/						
						/	/						
						/	/						
						/	/						
						/	/						
						/	/						
THIS RECORD IS CERTIFIED TRUE AND CORRECT PILOT'S SIGNATURE _____			PAGE TOTAL			/	/						
			PREVIOUS TOTAL			/	/						
			NEW TOTAL			/	/						

TYPE OF PILOTING TIME						REMARKS AND ENDORSEMENTS
PIC	SIC	AS FLIGHT INSTRUCTOR	DUAL RECEIVED			

| DATE (MON/DAY) | AIRCRAFT TYPE | AIRCRAFT IDENT | ROUTE OF FLIGHT | | FLIGHT NUMBER | TAKE OFF & LANDING | | | CONDITION OF FLIGHT | | | | FLIG SIMUL |
| | | | FROM | TO | | DAY | NIGHT | AUTO LAND | NIGHT | ACTUAL INSTRUMENT | APP | | |
											R/W NO.	TYPE	
						/	/						
						/	/						
						/	/						
						/	/						
						/	/						
						/	/						
						/	/						
						/	/						
						/	/						
						/	/						
						/	/						
THIS RECORD IS CERTIFIED TRUE AND CORRECT PILOT'S SIGNATURE _____			PAGE TOTAL			/	/						
			PREVIOUS TOTAL			/	/						
			NEW TOTAL			/	/						

TYPE OF PILOTING TIME						REMARKS AND ENDORSEMENTS
PIC	SIC	AS FLIGHT INSTRUCTOR	DUAL RECEIVED			

DATE (MON/DAY)	AIRCRAFT TYPE	AIRCRAFT IDENT	ROUTE OF FLIGHT		FLIGHT NUMBER	TAKE OFF & LANDING			CONDITION OF FLIGHT				FLIG SIMULA
			FROM	TO		DAY	NIGHT	AUTO LAND	NIGHT	ACTUAL INSTRUMENT	APP		
											R/W NO.	TYPE	
						/	/						
						/	/						
						/	/						
						/	/						
						/	/						
						/	/						
						/	/						
						/	/						
						/	/						
						/	/						
						/	/						
THIS RECORD IS CERTIFIED TRUE AND CORRECT PILOT'S SIGNATURE _____			PAGE TOTAL			/	/						
			PREVIOUS TOTAL			/	/						
			NEW TOTAL			/	/						

TYPE OF PILOTING TIME						REMARKS AND ENDORSEMENTS
PIC	SIC	AS FLIGHT INSTRUCTOR	DUAL RECEIVED			

DATE (MON/DAY)	AIRCRAFT TYPE	AIRCRAFT IDENT	ROUTE OF FLIGHT		FLIGHT NUMBER	TAKE OFF & LANDING			CONDITION OF FLIGHT				FLIG SIMUL.
			FROM	TO		DAY	NIGHT	AUTO LAND	NIGHT	ACTUAL INSTRUMENT	APP		
											R/W NO.	TYPE	
						/	/						
						/	/						
						/	/						
						/	/						
						/	/						
						/	/						
						/	/						
						/	/						
						/	/						
						/	/						
						/	/						

THIS RECORD IS CERTIFIED TRUE AND CORRECT PILOT'S SIGNATURE _____		PAGE TOTAL	/	/					
		PREVIOUS TOTAL	/	/					
		NEW TOTAL	/	/					

TYPE OF PILOTING TIME						REMARKS AND ENDORSEMENTS
PIC	SIC	AS FLIGHT INSTRUCTOR	DUAL RECEIVED			

DATE (MON/DAY)	AIRCRAFT TYPE	AIRCRAFT IDENT	ROUTE OF FLIGHT		FLIGHT NUMBER	TAKE OFF & LANDING			CONDITION OF FLIGHT				FLIG SIMUL.
			FROM	TO		DAY	NIGHT	AUTO LAND	NIGHT	ACTUAL INSTRUMENT	APP		
											R/W NO.	TYPE	
						/	/						
						/	/						
						/	/						
						/	/						
						/	/						
						/	/						
						/	/						
						/	/						
						/	/						
						/	/						
						/	/						
THIS RECORD IS CERTIFIED TRUE AND CORRECT PILOT'S SIGNATURE _____			PAGE TOTAL			/	/						
			PREVIOUS TOTAL			/	/						
			NEW TOTAL			/	/						

TYPE OF PILOTING TIME						REMARKS AND ENDORSEMENTS
PIC	SIC	AS FLIGHT INSTRUCTOR	DUAL RECEIVED			

DATE (MON/DAY)	AIRCRAFT TYPE	AIRCRAFT IDENT	ROUTE OF FLIGHT		FLIGHT NUMBER	TAKE OFF & LANDING			CONDITION OF FLIGHT				FLIGHT SIMULA
			FROM	TO		DAY	NIGHT	AUTO LAND	NIGHT	ACTUAL INSTRUMENT	APP		
											R/W NO.	TYPE	
						/	/						
						/	/						
						/	/						
						/	/						
						/	/						
						/	/						
						/	/						
						/	/						
						/	/						
						/	/						
						/	/						
THIS RECORD IS CERTIFIED TRUE AND CORRECT PILOT'S SIGNATURE _____			PAGE TOTAL			/	/						
			PREVIOUS TOTAL			/	/						
			NEW TOTAL			/	/						

TYPE OF PILOTING TIME						REMARKS AND ENDORSEMENTS
PIC	SIC	AS FLIGHT INSTRUCTOR	DUAL RECEIVED			

| DATE (MON/DAY) | AIRCRAFT TYPE | AIRCRAFT IDENT | ROUTE OF FLIGHT | | FLIGHT NUMBER | TAKE OFF & LANDING | | | CONDITION OF FLIGHT | | | | FLIG SIMULA |
| | | | FROM | TO | | DAY | NIGHT | AUTO LAND | NIGHT | ACTUAL INSTRUMENT | APP | | |
											R/W NO.	TYPE	
						/	/						
						/	/						
						/	/						
						/	/						
						/	/						
						/	/						
						/	/						
						/	/						
						/	/						
						/	/						
						/	/						
THIS RECORD IS CERTIFIED TRUE AND CORRECT PILOT'S SIGNATURE _____			PAGE TOTAL			/	/						
			PREVIOUS TOTAL			/	/						
			NEW TOTAL			/	/						

TYPE OF PILOTING TIME						REMARKS AND ENDORSEMENTS
PIC	SIC	AS FLIGHT INSTRUCTOR	DUAL RECEIVED			

DATE (MON/DAY)	AIRCRAFT TYPE	AIRCRAFT IDENT	ROUTE OF FLIGHT		FLIGHT NUMBER	TAKE OFF & LANDING			CONDITION OF FLIGHT				FLIG SIMUL
			FROM	TO		DAY	NIGHT	AUTO LAND	NIGHT	ACTUAL INSTRUMENT	APP		
											R/W NO.	TYPE	
						/	/						
						/	/						
						/	/						
						/	/						
						/	/						
						/	/						
						/	/						
						/	/						
						/	/						
						/	/						
						/	/						
THIS RECORD IS CERTIFIED TRUE AND CORRECT PILOT'S SIGNATURE _____			PAGE TOTAL			/	/						
			PREVIOUS TOTAL			/	/						
			NEW TOTAL			/	/						

TYPE OF PILOTING TIME						REMARKS AND ENDORSEMENTS
PIC	SIC	AS FLIGHT INSTRUCTOR	DUAL RECEIVED			

DATE (MON/DAY)	AIRCRAFT TYPE	AIRCRAFT IDENT	ROUTE OF FLIGHT		FLIGHT NUMBER	TAKE OFF & LANDING			CONDITION OF FLIGHT				FLIG SIMUL
			FROM	TO		DAY	NIGHT	AUTO LAND	NIGHT	ACTUAL INSTRUMENT	R/W NO.	TYPE	
						/	/						
						/	/						
						/	/						
						/	/						
						/	/						
						/	/						
						/	/						
						/	/						
						/	/						
						/	/						
						/	/						
THIS RECORD IS CERTIFIED TRUE AND CORRECT PILOT'S SIGNATURE _____			PAGE TOTAL			/	/						
			PREVIOUS TOTAL			/	/						
			NEW TOTAL			/	/						

TYPE OF PILOTING TIME						REMARKS AND ENDORSEMENTS
PIC	SIC	AS FLIGHT INSTRUCTOR	DUAL RECEIVED			

| DATE (MON/DAY) | AIRCRAFT TYPE | AIRCRAFT IDENT | ROUTE OF FLIGHT | | FLIGHT NUMBER | TAKE OFF & LANDING | | | CONDITION OF FLIGHT | | | | FLIG SIMUL |
| | | | FROM | TO | | DAY | NIGHT | AUTO LAND | NIGHT | ACTUAL INSTRUMENT | APP | | |
											R/W NO.	TYPE	
						/	/						
						/	/						
						/	/						
						/	/						
						/	/						
						/	/						
						/	/						
						/	/						
						/	/						
						/	/						
						/	/						
THIS RECORD IS CERTIFIED TRUE AND CORRECT PILOT'S SIGNATURE _____			PAGE TOTAL			/	/						
			PREVIOUS TOTAL			/	/						
			NEW TOTAL			/	/						

TYPE OF PILOTING TIME						REMARKS AND ENDORSEMENTS
PIC	SIC	AS FLIGHT INSTRUCTOR	DUAL RECEIVED			

DATE (MON/DAY)	AIRCRAFT TYPE	AIRCRAFT IDENT	ROUTE OF FLIGHT		FLIGHT NUMBER	TAKE OFF & LANDING			CONDITION OF FLIGHT				FLIGHT SIMULA
			FROM	TO		DAY	NIGHT	AUTO LAND	NIGHT	ACTUAL INSTRUMENT	APP		
											R/W NO.	TYPE	
						/	/						
						/	/						
						/	/						
						/	/						
						/	/						
						/	/						
						/	/						
						/	/						
						/	/						
						/	/						
						/	/						

THIS RECORD IS CERTIFIED TRUE AND CORRECT PILOT'S SIGNATURE _____

PAGE TOTAL	/	/
PREVIOUS TOTAL	/	/
NEW TOTAL	/	/

TYPE OF PILOTING TIME						REMARKS AND ENDORSEMENTS
PIC	SIC	AS FLIGHT INSTRUCTOR	DUAL RECEIVED			

DATE (MON/DAY)	AIRCRAFT TYPE	AIRCRAFT IDENT	ROUTE OF FLIGHT		FLIGHT NUMBER	TAKE OFF & LANDING			CONDITION OF FLIGHT				FLIG SIMUL
			FROM	TO		DAY	NIGHT	AUTO LAND	NIGHT	ACTUAL INSTRUMENT	APP R/W NO.	TYPE	
						/	/						
						/	/						
						/	/						
						/	/						
						/	/						
						/	/						
						/	/						
						/	/						
						/	/						
						/	/						
						/	/						
THIS RECORD IS CERTIFIED TRUE AND CORRECT PILOT'S SIGNATURE _____			PAGE TOTAL			/	/						
			PREVIOUS TOTAL			/	/						
			NEW TOTAL			/	/						

TYPE OF PILOTING TIME						REMARKS AND ENDORSEMENTS
PIC	SIC	AS FLIGHT INSTRUCTOR	DUAL RECEIVED			

DATE (MON/DAY)	AIRCRAFT TYPE	AIRCRAFT IDENT	ROUTE OF FLIGHT		FLIGHT NUMBER	TAKE OFF & LANDING			CONDITION OF FLIGHT				FLIGHT SIMULA
			FROM	TO		DAY	NIGHT	AUTO LAND	NIGHT	ACTUAL INSTRUMENT	APP		
											R/W NO.	TYPE	
						/	/						
						/	/						
						/	/						
						/	/						
						/	/						
						/	/						
						/	/						
						/	/						
						/	/						
						/	/						
						/	/						
THIS RECORD IS CERTIFIED TRUE AND CORRECT PILOT'S SIGNATURE _____			PAGE TOTAL			/	/						
			PREVIOUS TOTAL			/	/						
			NEW TOTAL			/	/						

TYPE OF PILOTING TIME						REMARKS AND ENDORSEMENTS
PIC	SIC	AS FLIGHT INSTRUCTOR	DUAL RECEIVED			

DATE (MON/DAY)	AIRCRAFT TYPE	AIRCRAFT IDENT	ROUTE OF FLIGHT		FLIGHT NUMBER	TAKE OFF & LANDING			CONDITION OF FLIGHT				FLIG SIMULA
			FROM	TO		DAY	NIGHT	AUTO LAND	NIGHT	ACTUAL INSTRUMENT	APP		
											R/W NO.	TYPE	
						/	/						
						/	/						
						/	/						
						/	/						
						/	/						
						/	/						
						/	/						
						/	/						
						/	/						
						/	/						
						/	/						
THIS RECORD IS CERTIFIED TRUE AND CORRECT PILOT'S SIGNATURE _____			PAGE TOTAL			/	/						
			PREVIOUS TOTAL			/	/						
			NEW TOTAL			/	/						

TYPE OF PILOTING TIME						REMARKS AND ENDORSEMENTS
PIC	SIC	AS FLIGHT INSTRUCTOR	DUAL RECEIVED			

DATE (MON/DAY)	AIRCRAFT TYPE	AIRCRAFT IDENT	ROUTE OF FLIGHT		FLIGHT NUMBER	TAKE OFF & LANDING			CONDITION OF FLIGHT				FLIG SIMUL/
			FROM	TO		DAY	NIGHT	AUTO LAND	NIGHT	ACTUAL INSTRUMENT	APP		
											R/W NO.	TYPE	
						/	/						
						/	/						
						/	/						
						/	/						
						/	/						
						/	/						
						/	/						
						/	/						
						/	/						
						/	/						
						/	/						
THIS RECORD IS CERTIFIED TRUE AND CORRECT PILOT'S SIGNATURE _____			PAGE TOTAL			/	/						
			PREVIOUS TOTAL			/	/						
			NEW TOTAL			/	/						

TYPE OF PILOTING TIME						REMARKS AND ENDORSEMENTS
PIC	SIC	AS FLIGHT INSTRUCTOR	DUAL RECEIVED			

| DATE (MON/DAY) | AIRCRAFT TYPE | AIRCRAFT IDENT | ROUTE OF FLIGHT | | FLIGHT NUMBER | TAKE OFF & LANDING | | | CONDITION OF FLIGHT | | | | FLIG SIMUL |
| | | | FROM | TO | | DAY | NIGHT | AUTO LAND | NIGHT | ACTUAL INSTRUMENT | APP | | |
											R/W NO.	TYPE	
						/	/						
						/	/						
						/	/						
						/	/						
						/	/						
						/	/						
						/	/						
						/	/						
						/	/						
						/	/						
						/	/						
THIS RECORD IS CERTIFIED TRUE AND CORRECT PILOT'S SIGNATURE _____			PAGE TOTAL			/	/						
			PREVIOUS TOTAL			/	/						
			NEW TOTAL			/	/						

TYPE OF PILOTING TIME						REMARKS AND ENDORSEMENTS
PIC	SIC	AS FLIGHT INSTRUCTOR	DUAL RECEIVED			

DATE (MON/DAY)	AIRCRAFT TYPE	AIRCRAFT IDENT	ROUTE OF FLIGHT		FLIGHT NUMBER	TAKE OFF & LANDING			CONDITION OF FLIGHT				FLIG SIMUL
			FROM	TO		DAY	NIGHT	AUTO LAND	NIGHT	ACTUAL INSTRUMENT	APP		
											R/W NO.	TYPE	
						/	/						
						/	/						
						/	/						
						/	/						
						/	/						
						/	/						
						/	/						
						/	/						
						/	/						
						/	/						
						/	/						
THIS RECORD IS CERTIFIED TRUE AND CORRECT PILOT'S SIGNATURE _____			PAGE TOTAL			/	/						
			PREVIOUS TOTAL			/	/						
			NEW TOTAL			/	/						

TYPE OF PILOTING TIME						REMARKS AND ENDORSEMENTS
PIC	SIC	AS FLIGHT INSTRUCTOR	DUAL RECEIVED			

DATE (MON/DAY)	AIRCRAFT TYPE	AIRCRAFT IDENT	ROUTE OF FLIGHT		FLIGHT NUMBER	TAKE OFF & LANDING			CONDITION OF FLIGHT				FLIG SIMUL
			FROM	TO		DAY	NIGHT	AUTO LAND	NIGHT	ACTUAL INSTRUMENT	R/W NO.	TYPE	
						/	/						
						/	/						
						/	/						
						/	/						
						/	/						
						/	/						
						/	/						
						/	/						
						/	/						
						/	/						
						/	/						
THIS RECORD IS CERTIFIED TRUE AND CORRECT PILOT'S SIGNATURE _____			PAGE TOTAL			/	/						
			PREVIOUS TOTAL			/	/						
			NEW TOTAL			/	/						

TYPE OF PILOTING TIME						REMARKS AND ENDORSEMENTS
PIC	SIC	AS FLIGHT INSTRUCTOR	DUAL RECEIVED			

DATE (MON/DAY)	AIRCRAFT TYPE	AIRCRAFT IDENT	ROUTE OF FLIGHT		FLIGHT NUMBER	TAKE OFF & LANDING			CONDITION OF FLIGHT				FLIG SIMUL
			FROM	TO		DAY	NIGHT	AUTO LAND	NIGHT	ACTUAL INSTRUMENT	APP		
											R/W NO.	TYPE	
						/	/						
						/	/						
						/	/						
						/	/						
						/	/						
						/	/						
						/	/						
						/	/						
						/	/						
						/	/						
						/	/						

THIS RECORD IS CERTIFIED TRUE AND CORRECT PILOT'S SIGNATURE _____			PAGE TOTAL	/	/					
			PREVIOUS TOTAL	/	/					
			NEW TOTAL	/	/					

TYPE OF PILOTING TIME						REMARKS AND ENDORSEMENTS
PIC	SIC	AS FLIGHT INSTRUCTOR	DUAL RECEIVED			

발행일 2019년 1월 13일 1판1쇄 발행   **발행처** 듀오북스   **펴낸이** 박승희   **펴낸이** 항공기술직업전문학교

등록일자 2018년 10월 12일 제2018-000281호   **주소** 서울시 마포구 환일2길 5-1

편집부 (070)7807_3690   **팩스** (050)4277_8651

웹사이트 www.duobooks.co.kr

정가 13,000원 **ISBN** 979-11-965450-3-1 10550